AF725325

ALGEBRA at School

It's Everyplace You Are!

Grade 3

By

Jill Osofsky

Published by Frank Schaffer Publications
an imprint of

McGraw Hill Children's Publishing

Author: Jill Osofsky
Editor: Jerry Aten

McGraw-Hill Children's Publishing

Published by Frank Schaffer Publications
An imprint of McGraw-Hill Children's Publishing

Send all inquiries to:
McGraw-Hill Children's Publishing
3195 Wilson Drive NW
Grand Rapids, Michigan 49544

Algebra at School: It's Everyplace You Are!—grade 3
ISBN: 0-7682-2511-6

1 2 3 4 5 6 7 8 9 PHX BK 07 06 05 04 03 02

The McGraw-Hill Companies

Algebra at School—It's Everyplace You Are
Grade 3

Table of Contents

Algebra at School—It's Everyplace You Are

The *It's Everyplace You Are* series makes school curriculum come alive. Engaging, hands-on activities help students see the connection between daily experiences and beginning algebra concepts. The activities are relevant to students' daily lives, drawing on familiar materials and experiences in the school environment.

This book includes six units to provide a variety of early algebra experiences.

- In the Classroom
- On the Playground
- On the Way to School
- In the Lunchroom
- In the Library
- In Other Places at School

Meeting the Standards

This book has been designed to meet the Standards for Mathematics defined by the National Council of Teachers of Mathematics (NCTM). The activities provide an informal introduction to algebra concepts through experiences in sorting and classifying, working with patterns to make generalizations, working with variables, and using logical reasoning to solve problems. While the activities are not labeled with the language of formal algebra, these experiences do lay the foundation for successful work with algebra concepts in higher grades.

The activity pages may be used in a variety of ways. Here are a few suggestions:

- Reproduce each activity as a transparency to guide the whole class.
- Duplicate the activity for each student or pair of students to solve.
- Duplicate the activity for a take-home project.

Materials

The activities call for the use of standard classroom materials or materials commonly found in the classroom and school. When introducing new concepts, use objects such as counters or cubes to allow children to work out the mathematical situations. Concrete, hands-on experiences are most appropriate at this level. Once the concept has been introduced and practiced on the concrete level, pictures and mathematical symbols can be used.

Assessment

Encourage children to talk about how they solved each problem, guiding them to see relationships and begin to make conjectures and generalizations.

Algebra in the Classroom

The classroom provides many opportunities to introduce beginning algebra concepts. This unit includes activities identifying, extending, and predicting number patterns. A T-chart is introduced as a tool used to organize patterns. Observations about the patterns develop the skills required for finding relationships between numbers and making generalizations. Early experiences with number patterns lay the foundation for understanding functions and variables work in higher grades.

Encourage students to describe the patterns in numbers. As they work on the T-charts ask questions such as *Can you think of a rule about this pattern that will work with all numbers or in other situations?* Emphasis should be placed on the number patterns and relationships identified, then writing equations to represent the relationship.

Prepare for Activities

✓ Model using a T-chart to organize patterns with the whole group. Invite students to identify and name the number pattern in the first activity. If students have difficulty recognizing the pattern elements try using concrete objects to introduce each pattern first.

✓ Provide concrete objects such as cubes or counters to act out number patterns as needed.

Concepts

- Patterns can be identified in numbers.
- A T-chart is a useful tool for organizing patterns.
- Rules or generalizations state how a pattern works in a numerical situation.
- Identifying number patterns helps to solve problems.

Extension Ideas

- Invite students to identify and organize patterns they find in the classroom. For example, is there a pattern in the arrangement of crayons in boxes of 48 and 64 crayons? What patterns can be found in the calendar? Is there a pattern in the number of homework problems assigned daily?
- Introduce cinquain or haiku poetry to the class. Challenge the students to look for patterns in the number of lines in each poetry form. Can they find a pattern in the number of syllables in each poetry form?
- Provide copies of a hundred chart and have children color in patterns for 2s, 5s, 3s, and other multiples. Encourage talk about the patterns. Repeat the activity with patterns for other numbers up to 12s.

Name________________________ Date________________________

Pairs in Chairs

There is a pattern in this classroom.

Each desk seats 2 students.

How many students are seated in 5 desks?

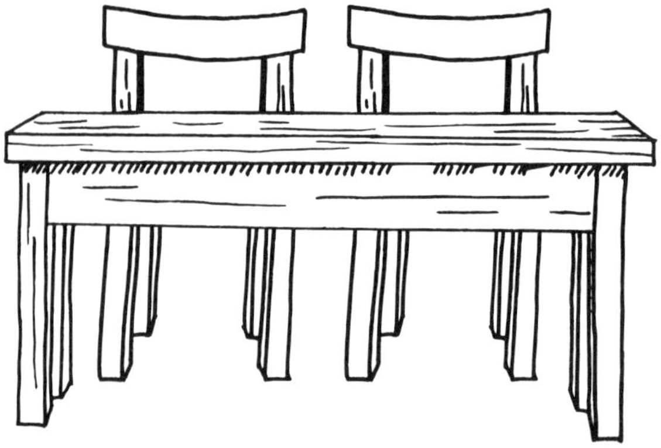

Directions:

Fill in the T-chart.

Look for a pattern.

It will help you solve the problem.

Number of desks	Number of students
1	2
2	4
3	6
4	______
5	______

Name the pattern. ________________________________

If the pattern continues in the same way, how many students in 6 desks? ________________________________

How many desks to seat pairs of students in your classroom? ________________________________

Name________________________ Date________________________

Perfect Papers

There's a pattern in this classroom.

The bulletin board has a display of perfect papers.

There are 5 papers in each row.

There are 5 rows of papers.

How many perfect papers in all?

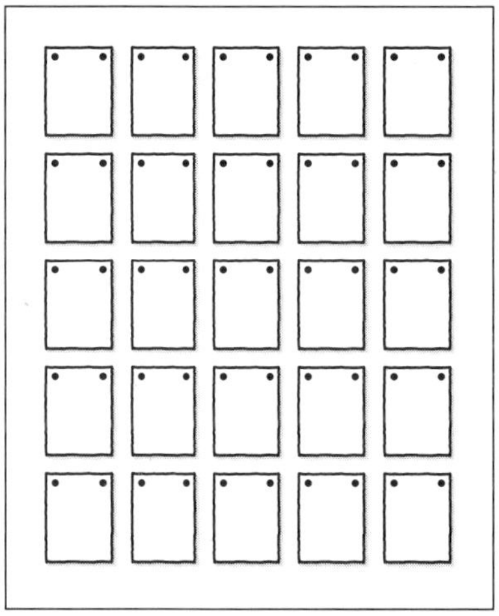

Directions

Fill in the T-chart.

Look for a pattern.

It will help you solve the problem.

Number of rows	Number of perfect papers
1	5
2	10
3	15
4	____
5	____

Name the pattern. ________________________

If the pattern continues in the same way, how many perfect papers in 6 rows? ________________________

Name______________________________ Date______________________________

Handling Handwriting

There's a pattern in this classroom.

The third graders are learning cursive handwriting.

They learn 4 new letters each week.

How many letters will they learn in 5 weeks?

Directions:

Fill in the T-chart.

Look for a pattern.

It will help you solve the problem.

abcd

Number of weeks	Number of new cursive letters
1	4
2	8
3	12
4	______
5	______

Name the pattern. ______________________________

If the pattern continues in the same way, how many weeks to learn 24 letters? ______________________

Name________________________ Date________________________

Recreational Reading

There's a pattern in this classroom.

Everyone reads every day right after recess.

Everyone reads for pleasure.

Everyone reads for 10 minutes.

How many minutes of recreational reading in 5 days?

Directions:

Fill in the T-chart.

Look for a pattern.

It will help you solve the problem.

Number of days	Number of minutes
1	10
2	20
3	30
4	____
5	____

Name the pattern. ________________________________

If the pattern continues in the same way, how many minutes will each student read in 10 days? ________________

Name________________________________ Date________________________________

Poetry Patterns

There's a pattern in this classroom.

The third graders are writing haiku poetry.

Each haiku poem has 3 lines.

How many lines in 5 haiku poems?

Directions:

Fill in the T-chart.

Look for a pattern.

It will help you solve the problem.

Number of haiku poems	Number of lines
1	3
2	6
3	9
4	______
5	______

Name the pattern. ________________________________

If the pattern continues in the same way, how many lines in 6 haiku poems? ________________________

Name______________________________ Date______________________________

New Spelling Words

There's a pattern in this classroom.

Each week the students learn 20 new spelling words.

How many new words will they learn in 5 weeks?

Spelling Words	
1.	11.
2.	12.
3.	13.
4.	14.
5.	15.
6.	16.
7.	17.
8.	18.
9.	19.
10.	20.

Directions:

Fill in the T-chart.

Look for a pattern.

It will help you solve the problem.

Week	Number of new words
1	20
2	40
3	60
4	_____
5	_____

Name the pattern. ______________________________

If the pattern continues in the same way,
how many new spelling words in 6 weeks? ______________________________

Algebra in the Lunchroom

The lunchroom provides many opportunities to introduce beginning algebra concepts. This unit includes informal experiences with number patterns and functions. The students observe the numbers in each table carefully to discover the relationship between the pairs of numbers. By observing the relationships between the number pairs students can see the rule or generalization that defines the relationship. The ability to state a rule or generalization based on observation lays the foundation for work with proportional relationships and functions and variables in higher grades.

Help children to look carefully at the numbers in each table to identify the relationship between the numbers. Invite them to talk about the patterns found in the tables. Some children will observe patterns in operations and relationships and between numbers, such as counting by 2s, 5s, and 10s. Ask questions such as *What pattern do you see in the numbers in the top row? Can you find a pattern in the numbers in the bottom row? Can you find a pattern in each row? What operation do you see here? Can you name a rule for the pattern?*

Prepare for Activities

✓ The activities feature using a secret rule for determining quantities of food needed to serve a meal in the cafeteria. Students are required to look for a pattern in the table between the top row and bottom row of numbers. Once the pattern has been identified, they use the secret rule to fill in the missing numbers and solve the problem.

✓ Make a transparency of the activity to model using each type of table.

✓ Provide objects such as cubes or counters to act out number situations as needed.

Concepts

- Number patterns can be organized in a table.
- Tables can be used to identify a number pattern and describe it in words.
- Number patterns can be used to identify a rule or generalization about sequences or pairs of numbers.
- Understanding the rule for a pattern allows one to extend the pattern indefinitely.

Extension Ideas

- Invite students to look for number relationships in other locations and situations in school such as games played on the playground.
- Students may enjoy working with function machines on the Internet. Go to your favorite education Web site and enter "function machine" in the search box. Your search will reward you with many activities and lessons that your students will enjoy!

Name______________________________ Date______________________________

Cook's Rule for Sandwiches

Find the cook's rule for how much of each ingredient is needed.

Directions:

Look at each number in the table.

Look for a pattern between the numbers.

Name the rule.

Grilled cheese sandwiches

Number of sandwiches	4	5	6	7	8
Pieces of cheese	6	7	8	___	___
Rule:					

Turkey sandwiches

Number of sandwiches	3	4	5	7	10
Pieces of turkey	7	8	9	___	___
Rule:					

0-7682-2511-6 *Algebra at School—It's Everyplace You Are*

Name________________________ Date________________________

Cook's Rule for Pizza

Find the cook's rule for how much of each ingredient is needed.

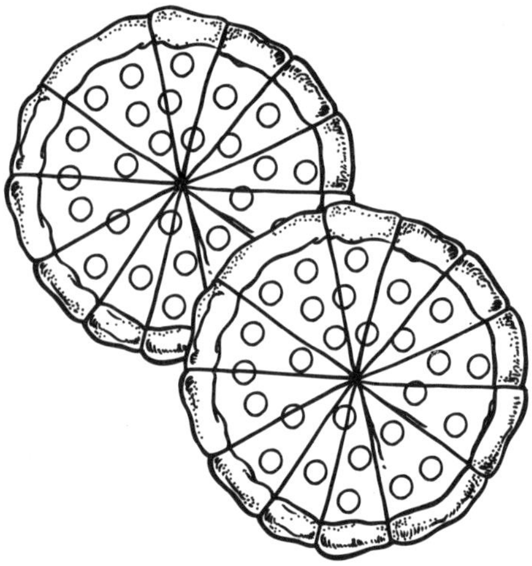

Directions:

Look at each number in the table.

Look for a pattern between the numbers.

Name the rule.

Pepperoni pizza

Number of slices	1	2	3	4	5
Pieces of pepperoni	6	7	8	___	___
Rule:					

Extra cheese pizza

Number of slices	3	4	5	7	10
Pieces of cheese	6	8	10	___	___
Rule:					

Name________________________ Date________________________

Cook's Rule for Spaghetti

Find the cook's rule for how much of each ingredient is needed.

Directions:

Look at each number in the table.

Look for a pattern between the numbers.

Name the rule.

Spaghetti with tomato sauce

Plates of spaghetti	6	7	9	10	12
Spoons of sauce	3	4	6	___	___
Rule:					

Spaghetti and meatballs

Plates of spaghetti	2	3	4	6	9
Number of meatballs	9	10	___	___	16
Rule:					

Name________________________________ Date________________________________

Cook's Rule for Burritos

Find the cook's rule for how much of each ingredient is needed.

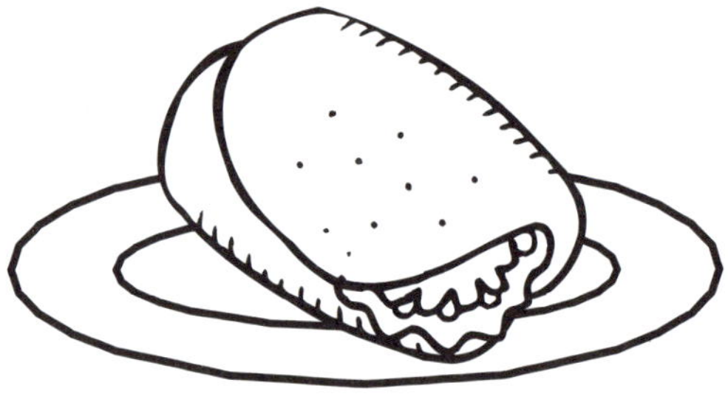

Directions:

Look at each number in the table.

Look for a pattern between the numbers.

Name the rule.

Bean burrito

Number of burritos	1	2	3	4	5
Spoons of beans	3	6	9	____	____
Rule:					

Meat burrito

Number of burritos	4	6	8	10	12
Spoons of meat	2	3	4	____	____
Rule:					

Name______________________ Date______________________

Cook's Rule for Nachos

Find the cook's rule for how much of each ingredient is needed.

Directions:

Look at each number in the table.

Look for a pattern between the numbers.

Name the rule.

Cheese nachos

Plate of nachos	3	4	5	6	9
Cups of sauce	9	12	___	___	___
Rule:					

Nachos supreme

Plate of nachos	4	5	6	10	20
Number of olives	8	10	___	___	___
Rule:					

Name________________________ Date________________________

Cook's Rule for Chocolate Chip Cookies

Find the cook's rule for how much of each ingredient is needed.

Directions:

Look at each number in the table.

Look for a pattern between the numbers.

Name the rule.

Plain chocolate chip cookies

Number of cookies	9	10	11	12	15
Number of chocolate chips	45	50	55	____	____
Rule:					

Chocolate chip cookies with nuts

Number of cookies	6	8	10	11	12
Number of nuts	18	24	____	____	____
Rule:					

Algebra on the Playground

The playground provides many opportunities to introduce beginning algebra concepts. This unit includes activities designed to analyze change using line graphs. Representing data to show change is a skill that students will use in other areas of the curriculum. Understanding how a line graph indicates change lays the foundation for more focused study of how the slope of a line represents the rate of change and algebraic reasoning in higher grades.

The President's Challenge for Physical Activity and Fitness provides a real, meaningful context for students analyzing change. The activities use a line graph to provide data on student performance in the four events. Invite students to describe the change and what may have caused the change. Ask questions such as: *What does the information show? How does the line indicate change? Can you find a pattern? What could have caused the change?*

Prepare for Activities

✓ Make a transparency of the activities to model using the line graphs. Point out the title, scale, and labels. Model how to identify each point on the line graph, first by locating the information on the horizontal axis and then by moving up vertically to where the two lines meet at the point. Point out that patterns of change can occur at a constant rate, such as someone running at a constant speed, or at rates that increase or decrease, as in the rate of plant growth.

Concepts

- Change can occur at a constant rate.
- Change can increase or decrease over time.
- Change can follow a pattern.
- A line graph can be used to show how change takes place over time.
- A point on a line graph relates to two related pieces of data, such as height and age.
- A line connects the points on a line graph.
- A line graph can be used to answer questions about change.

Extension Ideas

- If your school is not participating in the President's Challenge, you may wish to get more information on the activities. Go to www.fitness.gov. for more information.
- Look for opportunities to use tables and line graphs to represent change in other curriculum areas. You can have students observe and record daily temperatures over one month, graphing the change daily. Have students record the number of pages they read daily over a week's time, then record their times in a table or chart. Discuss the possible reasons for change and differences in the rate of change.
- Look for other situations in school to use for studies in change. For example, compare the number of hot lunches purchased over the course of a week. Discuss why the numbers change over the course of the week.

Name________________________________ Date________________________________

Curl-Up for Fitness

The president wants all Americans to be fit and healthy!

One third grader practiced curl-ups for one week.

How did the number of curl-ups completed change over the week?

Directions:

Look at the labels on the line graph.

Look at the points on the graph.

Use the information to answer each question.

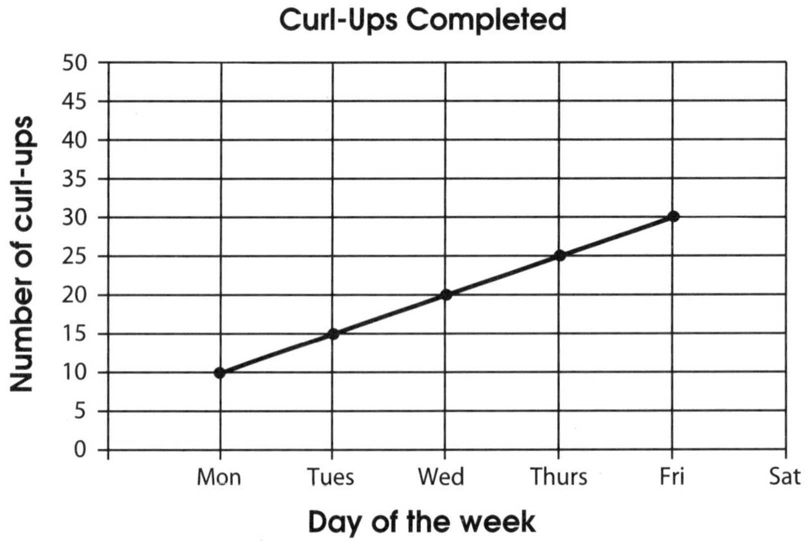

1. How many curl-ups did the third grader complete on Wednesday? ________________

2. What day of the week did the third grader complete 25 curl-ups? ________________

3. If the pattern continues, how many curl-ups will the student complete on Saturday? ________________

Name________________________ Date________________________

Shuttle Run for Fitness

The president wants all Americans to be fit and healthy!

One third grader practiced the shuttle run for one week.

How did the third grader's speed change over the week?

Directions:

Look at the labels on the line graph.

Look at the points on the graph.

Use the information to answer each question.

Shuttle Run Times

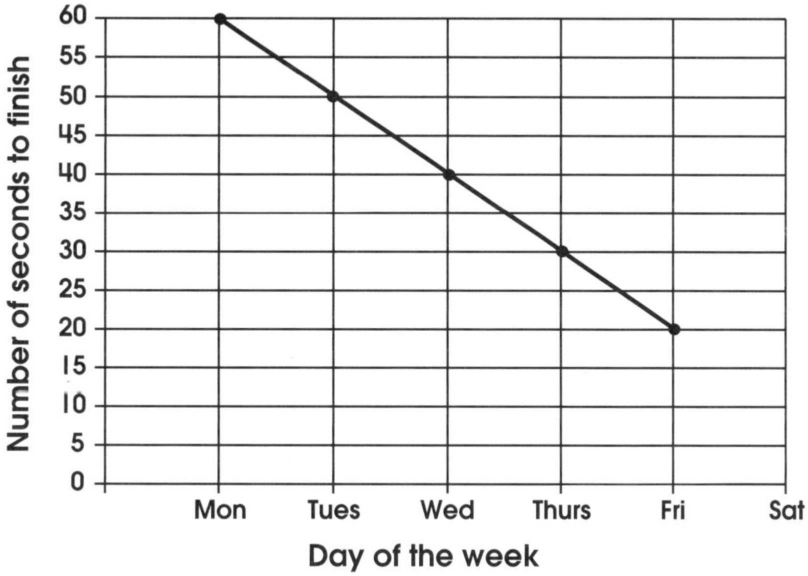

1. How many seconds to finish the shuttle run on Tuesday? ____________________

2. What day of the week did the third grader complete the run in 30 seconds? ____________________

3. If the pattern continues, how long to complete the shuttle run on Saturday? ____________________

Name______________________________ Date______________________________

Sit-and-Reach for Fitness

The president wants all Americans to be fit and healthy!

One third grader practiced the sit-and-reach for one week.

How did the length of the student's stretch change over the week?

Directions:

Look at the labels on the line graph.

Look at the points on the graph.

Use the information to answer each question.

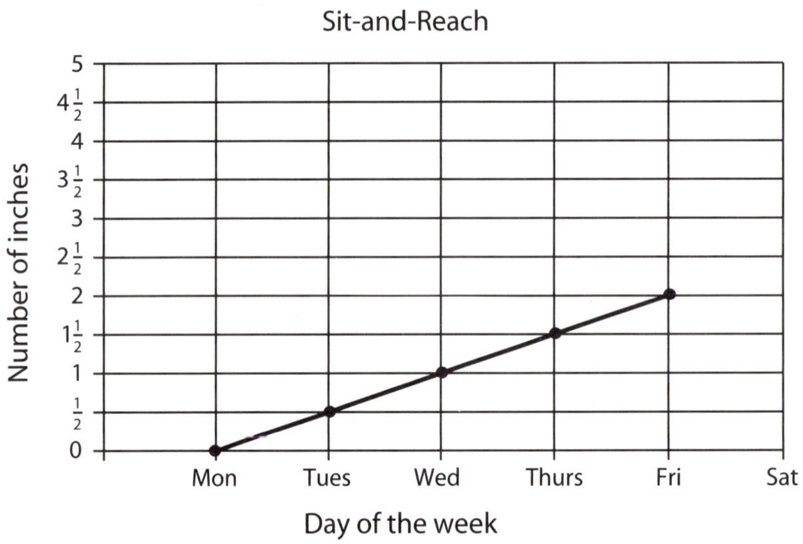

1. How far did the student reach on Monday? ____________________

2. What day of the week did the third grader reach 1 inch? ____________________

3. If the pattern continues in the same way, how far will the student reach on Saturday? ____________________

Name________________________ Date________________________

Make a Run for Fitness

The president wants all Americans to be fit and healthy!

One third grader practiced the half-mile run for one week.

How did the time taken change over the week?

Directions:

Look at the labels on the line graph.

Look at the points on the graph.

Use the information to answer each question.

1. How long did the run take on Monday? ____________________

2. What day of the week did the third grader complete the run in 5 minutes? ____________________

3. If the pattern continues in the same way, what day will the student complete the run in 4 minutes? ____________________

Name______________________________ Date______________________________

Pull-Up for Fitness

The president wants all Americans to be fit and healthy!

One third grader practiced pull-ups for one week.

How did the number of pull-ups change over the week?

Directions:

Look at the labels on the line graph.

Look at the points on the graph.

Use the information to answer each question.

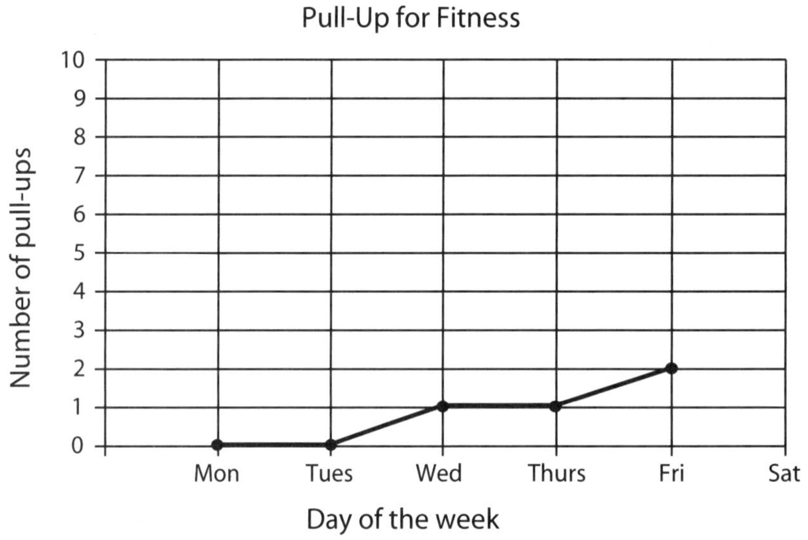

1. How many pull-ups did the student complete on Monday? ______________________________

2. What days of the week did the student complete 1 pull-up? ______________________________

3. If the pattern continues in the same way, how many pull-ups will the student complete on Saturday? ____________

Name________________________________ Date________________________________

The Healthy Lifestyle Award

The president wants all Americans to be fit and healthy!

The Active Lifestyle Award recognizes children who stay active at least 1 hour, 5 days a week, for 6 weeks.

Will this third grader win the Active Lifestyle Award?

Directions:

Look at the labels on the line graph.

Look at the points on the graph.

Use the information to answer each question.

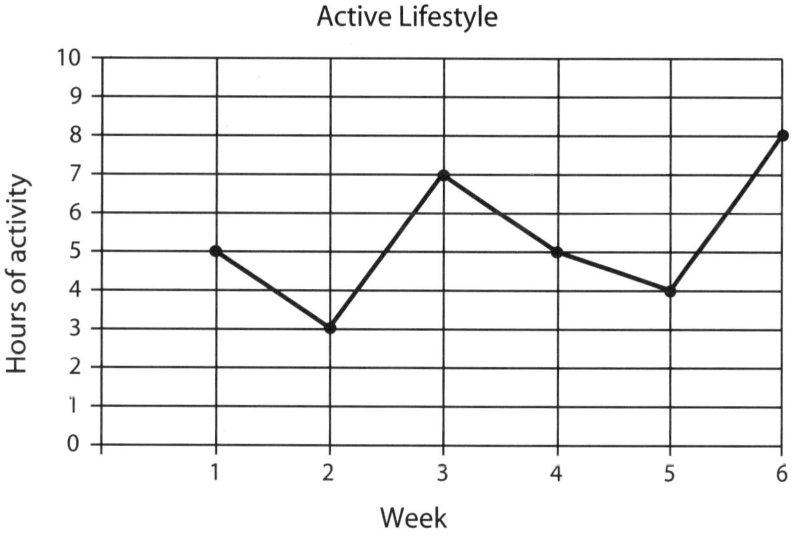

1. How many hours of activity in week 2? ____________________

2. How many hours of activity in week 4? ____________________

3. How many hours of activity in all 6 weeks? ____________________

4. Will the student win the award? ____________________

Algebra in the Library

The library provides many opportunities to introduce beginning algebra concepts. This unit includes activities working with growing patterns. Growing patterns involve an increase from one step to the next. Growing patterns follow some rule or generalization that allows one to predict what the pattern will be at any point in a sequence. Work with growing patterns lays the foundation for understanding the concept of function in higher algebra.

When discussing growing patterns, guide students to observe how each new sequence in the pattern "grows" from the previous one. For example, the stair step pattern "grows" by adding a row or a column to the previous shape. Ask questions such as *Can you tell me how you extended the pattern? How are growing patterns similar to other patterns you have worked with? How are they different? Can you think of a rule to describe the way the pattern grows?*

Prepare for Activities

✓ Building the patterns with objects helps students to identify and understand the concept of growing patterns. Provide concrete objects such as cubes or counters to build the patterns as needed.

Concepts

- Growing patterns involve a progression in size.
- Growing patterns feature a variable.
- Growing patterns can be organized in a table to predict the next member of a set.

Extension Ideas

- Challenge students to use pattern blocks to create physical models of growing patterns. Examples of shapes include a fish made from one hexagon and two triangles or an ice cream cone made from one trapezoid, one diamond, and two triangles. Students will enjoy making their own patterns to explore.
- Students will enjoy finding out more about another type of pattern, the Fibonnaci sequence. Each number in the Fibonnaci sequence, beginning with 2, is the sum of the two numbers preceding it in the series. This sequence occurs in the spiral rows on a pinecone and in the center of a daisy.

Name____________________ Date____________________

The Library Set-Up

The librarian used this pattern to group the bookcases in the library.

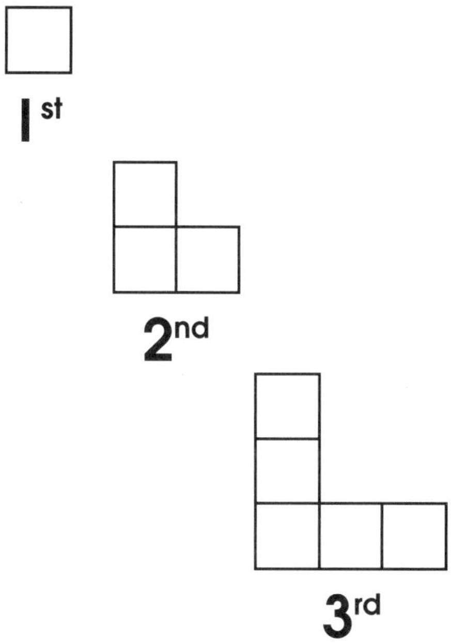

If the pattern continues in the same way,

how many bookcases in the next two groups?

Directions:

Fill in the table.

Look for a pattern in the numbers.

It can help you solve the problem.

Group	**1st**	**2nd**	**3rd**	**4th**	**5th**
Number of bookcases	**1**	**3**	**5**	____	____

Describe the pattern. ______________________________

Name________________________ Date________________________

The Book Fair Is Coming

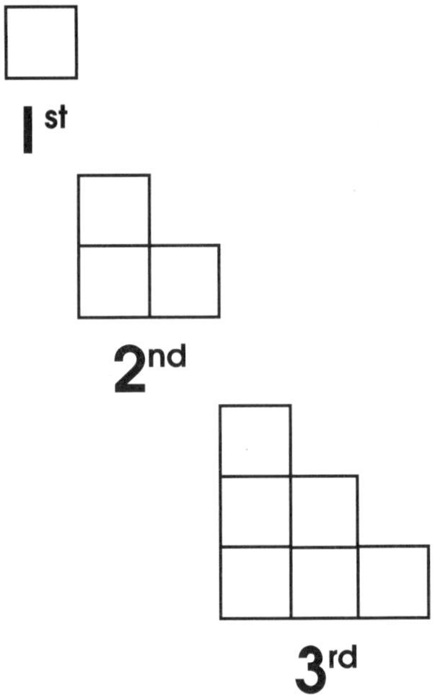

The library stores boxes of books for the book fair.

The boxes are stacked in a staircase pattern.

If the pattern continues in the same way,
how many boxes in the next two stacks?

Directions:

Fill in the table.

Look for a pattern in the numbers.

It can help you solve the problem.

Group	**1st**	**2nd**	**3rd**	**4th**	**5th**
Number of bookcases	**1**	**3**	**6**	____	____

Describe the pattern. ______________________________

Name________________________ Date________________________

Award-Winning Books

The librarian is making a display of Caldecott and Newbery books. She adds books to the display each day.

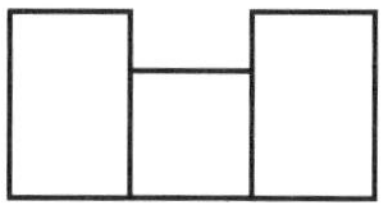

Day 1

If the pattern continues in the same way, how many books on display the fifth day?

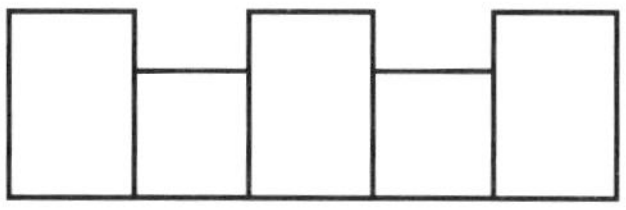

Day 2

Directions:

Fill in the table.

Look for a pattern in the numbers.

It can help you solve the problem.

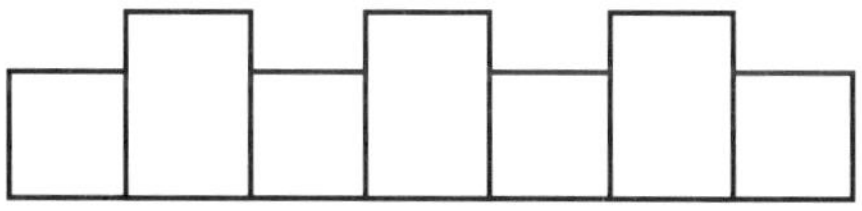

Day 3

Day	**1**	**2**	**3**	**4**	**5**
Number of books	**3**	**5**	**7**	____	____

Describe the pattern. ______________________________

Name________________________ Date________________________

Favorite Authors

The librarian put up posters of favorite books and authors.

A poster for 1 author shows 8 books.

A poster for 2 authors shows 10 books.

A poster for 3 authors shows 12 books.

If the pattern continues in the same way, how many books on a poster for 5 authors?

Directions:

Fill in the table.

Look for a pattern in the numbers.

It can help you solve the problem.

Number of authors	**1**	**2**	**3**	**4**	**5**
Number of books	**8**	**10**	**12**	____	____

Describe the pattern. ________________________________

Name________________________ Date________________________

Most Popular Books

The librarian has displays of most popular books.

The first display has 1 book.

The second display has 4 books.

The third display has 9 books.

If the pattern continues in the same way,
how many books on the fith display?

Directions:

Fill in the table.

Look for a pattern in the numbers.

It can help you solve the problem.

Display of authors	**1st**	**2nd**	**3rd**	**4th**	**5th**
Number of books	**1**	**4**	**9**	____	____

Describe the pattern. ________________________________

Name________________________________ Date________________________________

That Book Is Due

Students return library books by the due date.

The librarian collects the books each day.

On Monday, 1 book was returned.

On Tuesday, 3 books were returned.

On Wednesday, 6 books were returned.

If the pattern continues in the same way, how many books will be returned on Friday?

Directions:

Fill in the table.

Look for a pattern in the numbers.

It can help you solve the problem.

Day	**Monday (Day 1)**	**Tuesday (Day 2)**	**Wednesday (Day 3)**	**Thursday (Day 4)**	**Friday (Day 5)**
Number of books	**1**	**3**	**6**	______	______

Describe the pattern. ____

0-7682-2511-6 *Algebra at School—It's Everyplace You Are*

Algebra on the Way to School

Situations students encounter on the way to school provide many opportunities to introduce beginning algebra concepts. This unit includes activities finding variables. A variable represents an unknown quantity.

Students at this level should begin use a variety of strategies for finding unknown values. These include make a table and guess and check. The first four activities in this section use the guess-and-check strategy. The next two activities incorporate the look for a number pattern and make a table strategies introduced in earlier activities to find the unknown numbers. When working with the activities ask questions to guide student thinking. Ask questions such as *What number do I need to find? What numbers do I know? Do you think I should add or subtract? Who can think of a number I might try? What number pairs could I use to make equation true?*

Concepts

- The guess-and-check strategy can help to find unknown values.
- Numbers organized in a table or a list help to identify patterns.
- Identifying patterns can help solve for unknown values (variables).

Extension Ideas

- Have students write number sentences for each problem using a box, letter, or symbol to represent the unknown values.
- Magic square puzzles are fun for students and provide great practice finding unknown numbers. Find magic squares at *http://mathforum.org/alejandre/magic.square.html.*

Prepare for Activities

✓ The guess-and-check strategy may be new to some students. Model how to use the guess-and-check strategy to identify missing values to solve problems. Use the following story problem on a sheet of chart paper or an overhead transparency:

✓ Nancy has a bird feeder in her front yard. She counts 16 birds in all. There are three more robins than blackbirds. There are two fewer sparrows than blackbirds. How many of each kind of bird is at the bird feeder?

✓ After reading the problem together, ask volunteers to come up and make a guess for each of the missing values. Ask the class to use those values to check if the solutions are possible given the information in the problem. If the values are not correct, ask a volunteer to use the information from the incorrect guess to make another more informed guess. For example:

Guess: 3 blackbirds, 6 robins, 1 sparrow

3 + 6 + 1 = 10 The guess is too low

Guess: 5 blackbirds, 8 robins, 3 sparrows

5 + 8 + 3 = 16

✓ Provide concrete objects such as cubes or counters to act out number situations as needed.

Name_______________________ Date_______________________

Backpacks, Rolling Packs, and Book Bags

Third graders carry their school supplies in backpacks, wheeled backpacks, and book bags.

There are 12 third graders walking to school.

There are 2 more backpacks than wheeled backpacks.

There are 2 fewer book bags than wheeled backpacks.

How many of each kind?

Directions:

Use the guess-and-check strategy.

Make a guess for wheeled backpacks.

Check your guess.

Use the information from an incorrect guess to make another guess.

Check again.

Backpacks _______________

Wheeled backpacks _______________

Book bags _______________

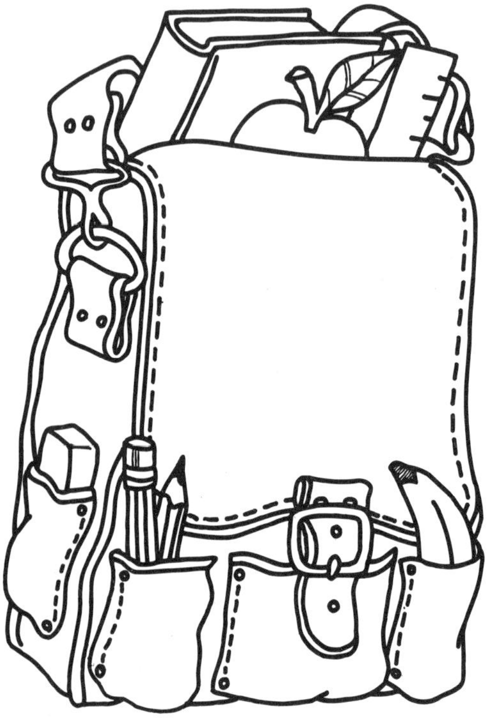

Name______________________________ Date______________________________

In the Place Where You Live

Joel passes by the homes of 18 friends on his way to school.

Some live in houses, some live in apartments, and some live in condominiums.

Three more friends live in apartments than condominiums.

Twice as many live in condominiums than in houses.

How many friends live in each of kind of home?

Directions:

Use the guess-and-check strategy.

Make a guess for condominiums.

Check your guess.

Use the information from an incorrect guess to make another guess.

Check again.

Houses ______________

Condominiums ______________

Apartments ______________

Name________________________________ Date________________________________

How Do You Get There?

The third graders are on their way to school.

Some are walking, some are riding in cars, and some are riding the bus.

There are 24 third graders.

Fewer walk to school than ride in a car.

Twice as many students ride in a car than walk.

Four more ride the bus than ride in a car.

How many students travel in each way?

Directions:

Use the guess-and-check strategy.

Make a guess for bus riders.

Check your guess.

Use the information from an incorrect guess to make another guess.

Check again.

Walk ______________

Ride in car ______________

Ride the bus ______________

Name________________________ Date________________________

Favorite Flags

People in Rachel's neighborhood fly flags.

Some fly American flags, some fly flags of the season, and some fly flags showing their favorite animal.

There are 24 flags in all.

The number of American flags and animal flags is the same.

There are three fewer season flags than American flags.

How many of each kind of flag?

Directions:

Use the guess-and-check strategy.

Make a guess for seasonal flags.

Check your guess.

Use the information from an incorrect guess to make another guess.

Check again.

American flags ______

Flags of the season ______

Animal flags ________

Name____________________ Date____________________

At the Crosswalk

Eighteen third graders are waiting at the crosswalk.

Some are boys and some are girls.

How many boys and how many girls could there be?

Find 5 ways to make groups of 18.

Directions:

Fill in the table.

Look for a pattern.

It will help you solve the problem.

	Way 1	Way 2	Way 3	Way 4	Way 5
Girls	3	____	5	____	7
Boys	15	14	____	12	____
Total	18	18	18	18	18

Name________________________ Date________________________

The Kids on the Bus

Here comes the school bus.

There are 24 third graders riding in all.

Some are boys and some are girls.

How many boys and how many girls could be on the bus.

Show 5 ways to make groups of 24.

Directions:

Fill in the table.

Look for a pattern.

It will help you solve the problem.

	Way 1	Way 2	Way 3	Way 4	Way 5
Girls	10	____	14	____	18
Boys	14	12	____	8	____
Total	24	24	24	24	24

Algebra in Other Places at School

Other locations around school provide many opportunities to introduce beginning algebra concepts. This unit includes activities practicing the concepts introduced in earlier sections of the book: identifying number patterns, organizing patterns in T-charts and tables, using graphs to analyze change, and working with variables to find unknown values.

Prepare for Activities

✓ Provide concrete objects such as cubes or counters to act out number situations as needed.

Concepts

- Patterns can be identified in numbers.
- Identifying number patterns helps to solve problems.
- A table or chart can be used to organize information.
- A line graph can be used to show changes over time.
- The guess-and-check strategy can help to find unknown values.
- Using objects to represent missing values helps to solve problems.
- Using logical reasoning helps to find missing values or numbers.

Extension Ideas

- Invite the children to make up their own activities about different areas in the school.
- Take the class for a walk around the school to look for algebra activities in your school.
- Hold a bake sale, penny drive, or math night at your school. Make up algebra activities to go with the numbers, patterns, and variables that make up your event!

Name______________________________ Date______________________________

A Penny Drive

The school is collecting pennies to raise money for field trips.

On Monday, 25 pennies were collected.

On Tuesday, 50 pennies were collected.

On Wednesday, 100 pennies were collected.

If the pattern continues on the same way, how many pennies will be collected on Friday?

Directions:

Use T-chart.

Look for a pattern.

It can help you solve the problem.

Day of the week	Number of classes
Monday	25
Tuesday	50
Wednesday	100
Thursday	______
Friday	______

How many pennies collected in all? ______________________________

Name________________________________ Date________________________________

The Pencil Machine

The student council has a pencil machine in front of school.

Pencils are sold to raise money for student activities.

How did the number of pencils sold change over the week?

Directions:

Look at the labels on the line graph.

Look at the points on the graph.

Use the information to answer each question.

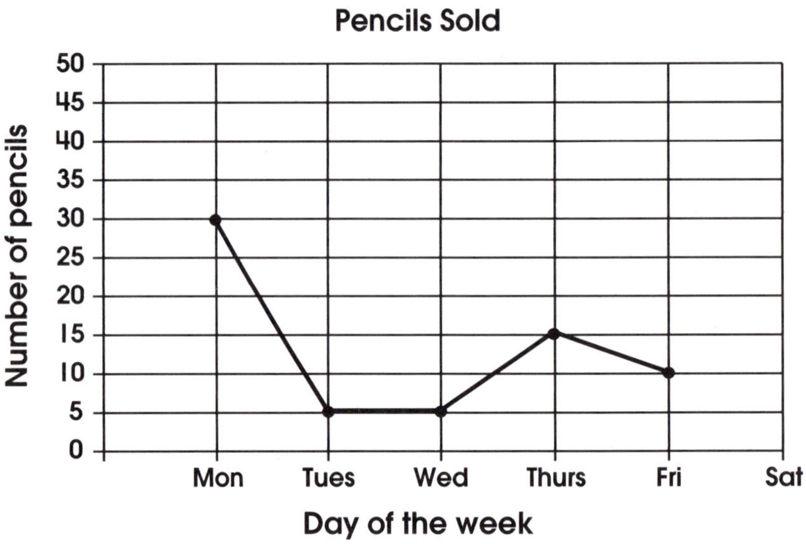

1. How many pencils were sold on Monday? ____________________

2. On what days of the week were the fewest pencils sold? ____________________

3. What might be the reason for the number of pencils sold on Monday? ____________________

Name________________________________ Date________________________________

Bake Sale

On Friday the student council had a bake sale.

They sold brownies, cookies, and cupcakes.

Three dozen baked goods were sold.

Twice as many cookies were sold than brownies.

Four more cupcakes than brownies were sold.

How many of each baked good was sold?

Directions:

Use the guess-and-check strategy.

Make a guess for brownies.

Check your guess.

Use the information from an incorrect guess to make another guess.

Check again.

Brownies __________

Cookies__________

Cupcakes ____________

Name________________________ Date________________________

Frenzied Frogs

The computer lab has a fun new math game, Frenzied Frogs.

When a student clicks on a number,

a different number of frogs hop across the screen.

Directions:

Look at each number in the table.

Look for a pattern between the numbers.

Name the rule for each game.

Number clicked	**12**	**8**	**10**	**20**	**18**
Number of frogs	**6**	**4**	**5**	____	____
Rule:					

Name______________________________ Date______________________________

In the Music Room

It's time to practice for the school music program.

There are 15 third graders practicing their instruments.

Twice as many students play the recorder than the violin.

Fewer students play the piano than the violin.

How many students play each kind of instrument?

Directions:

Use the guess-and-check strategy.

Make a guess for violins.

Check your guess.

Use the information from an incorrect guess to make another guess.

Check again.

Recorder ________

Violin __________

Piano _________

Name________________________ Date________________________

Math Night

It's Math Night in the school cafeteria.

Thirty families are playing games.

There are 4 games to choose from:

Pattern Puzzles, Coordinate Capers, What's My Rule, and Magic Squares.

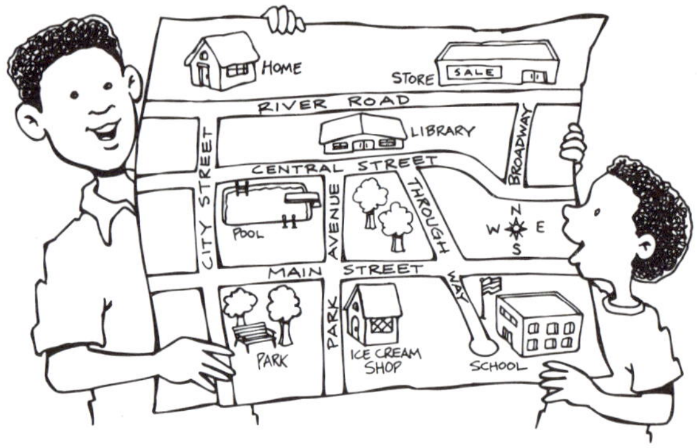

How many families could there be playing each game?

Show 5 ways to make groups of 30.

Directions:

Fill in the table.

Look for a pattern.

It will help you solve the problem.

	Way 1	Way 2	Way 3	Way 4	Way 5
Pattern Puzzles	10	8	7	____	6
Coordinate Capers	10	____	7	____	____
What's My Rule?	____	6	____	10	____
Magic Squares	5	6	____	5	____
Total	30	30	30	30	30

Algebra Vocabulary

function: number relationships in which each member of one set is paired with exactly one member of a second set

generalization: a rule that can be stated, often based on observing patterns

growing patterns: a pattern that contains a core sequence that increases from one set to the next

line graph: a graph connects points to show how data changes over time; used when numeric values are shown along a continuous number scale made of equally spaced points

pattern: something that repeats over and over in the same way; patterns can be found in colors, shapes, sizes, and numbers

T-chart: a chart for recording patterns

variable: a letter, picture, or symbol that represents an unknown quantity

Answer Key

Pairs in Chairs 6
8 students; 10 students
Pattern: Add 2 or counting by 2s; 12 students; answers will vary

Perfect Papers 7
20 papers; 25 papers
Pattern: Add 5 or counting by 5s; 30 papers

Handling Handwriting 8
16 letters; 20 letters
Pattern: Add 4 or counting by 4s; 6 weeks

Recreational Reading 9
40 minutes; 50 minutes
Pattern: Add 10 or counting by 10s; 100 pages

Poetry Patterns 10
12 lines; 15 lines
Pattern: Add 3 or counting by 3s; 18 lines

New Spelling Words 11
80 words; 100 words
Pattern: Add 20 or counting by 20s; 120 words

Cook's Rule for Sandwiches 13
14; 16. Rule: Add 2
11; 14. Rule: Add 4

Cook's Rule for Pizza 14
9; 14. Rule: Add 5
14; 20. Rule: Double the number of slices

Cook's Rule for Spaghetti 15
7; 9. Rule: Subtract 3
11;13. Rule: Add 7

Cook's Rule for Burritos 16
12; 15. Rule: Multiply number of burritos by 3 or repeat the number of burritos 3 times
5; 6. Rule: Halve the number of burritos or divide the number of burritos by 2

Cook's Rule for Nachos 17
15; 18; 27. Rule: Multiply by 3 or add the number of plates of nachos 3 times
12; 20; 40. Rule: Multiply by 2 or double the number of nachos

Cook's Rule for Chocolate Chip Cookies 18
60; 75. Rule: Multiply by 5 or add the number of cookies 5 times
30, 33, 36. Rule: Multiply by 3 or add the number of cookies 3 times

Curl-Up for Fitness 20
20; Thursday; 35

Shuttle Run for Fitness 21
50; Thursday; 10

Sit-and-Reach for Fitness 22
Zero inches; Wednesday; $2\frac{1}{2}$ inches

Make a Run for Fitness 23
9 minutes; Friday; Saturday

Pull-Up for Fitness 24
Zero; Wednesday and Thursday; 2

The Healthy Lifestyle Award 25
3; 5; 32 hours; the student will win the award

The Library Set-Up 27
7; 9
Pattern: The number of bookcases increases by 1 on each end or 2 book cases per group

The Book Fair Is Coming 28
10; 15
Pattern: Each row has one less box than the row below.

Award-Winning Books 29
9. 11
Pattern: Add 1 to each end

Favorite Authors 30
14; 16
Pattern: Two times the number of authors plus 6

Most Popular Books 31
16; 25
Pattern: Number in top row times itself

That Book Is Due 32
10; 15
Rule: Add the previous day's total to the number of each day

Backpacks, Rolling Packs, and Book Bags 34
6 backpacks; 4 wheeled backpacks; 2 book bags

In the Place Where You Live 35
Houses = 3; condominiums = 6; apartments = 9

How Do You Get There? 36
Walk = 4; car = 8; bus = 12

Favorite Flags 37
American flags = 9; season flags = 6; animal flags = 9

At the Crosswalk 38

	Way 1	Way 2	Way 3	Way 4	Way 5
Girls	3	4	5	6	7
Boys	15	14	13	12	11
Total	18	18	18	18	18

The Kids on the Bus 39

	Way 1	Way 2	Way 3	Way 4	Way 5
Girls	10	12	14	16	18
Boys	14	12	10	8	6
Total	24	24	24	24	24

A Penny Drive 41
200 pennies; 400 pennies; 775 pennies in all

The Pencil Machine 42
30 pencils; Tuesday and Wednesday; more students may have forgotten to bring pencils on Monday

Bake Sale 43
Brownies = 8; cookies = 16; cupcakes = 12

Frenzied Frogs 44
10; 9. Rule: Divide the number by 2, or one-half the number

In the Music Room 45
Recorder = 8; violin = 4; piano = 3

Math Night 46

	Way 1	Way 2	Way 3	Way 4	Way 5
Pattern Puzzles	10	8	7	5	6
Coordinate Capers	10	10	7	10	7
What's My Rule?	5	6	8	10	8
Magic Squares	5	6	8	5	9
Total	30	30	30	30	30